A Stepping-Stone Book

MATTER ALL AROUND YOU

A Book about Solids, Liquids, & Gases

By R. J. LEFKOWITZ

Illustrated by LAWRENCE DI FIORI

Parents' Magazine Press • New York

For Micah and Darien,
who really matter

Printed in the United States of America

Library of Congress Cataloging in Publication Data
Lefkowitz, R. J. 1942-
Matter all around you.
(A Stepping-stone book)
SUMMARY: Discusses the characteristics of the three states of matter—solids, liquids, and gases.
1. Matter—Juvenile literature. [1. Matter]
I. Di Fiori, Lawrence, illus. II. Title.
PZ10.L42Mat 530.4 73-179363
ISBN 0-8193-0550-2 (lib. bdg.)

What's In This Book

One

About Space and Things

All around you there is space.

Space to run in. Space to jump in. Space to walk, fly, sing, and laugh in.

Things take up some of the space. A rock takes up space. A bicycle takes up space. An eggplant takes up space.

You take up space.

Some things take up more space than others. A horse takes up more space than an eggplant. A car takes up more space than a bicycle. You take up more space than a baby brother or sister. But everything takes up *some* space.

Everything. Except a smile, a dream, a hope, a giggle. Things you think, things you say, things you dream or feel inside you, don't take up space.

Things you can touch do take up space. You can pet a horse or hold a rock in your hand. Can you hold a hope in your hands, or touch a giggle?

Things you can touch – things that take up space – are called *matter.*

Two

Matter All Around You

All around you there is matter.

A building is matter. A tree is matter. A book is matter. They take up space. If they are near you, you can touch them.

You are matter. Is your shadow matter? Can you touch it? You may be able to touch the place on the ground or on a wall where your shadow seems to be. But then the shadow seems to be on your finger. Can you feel it there?

You cannot really touch your shadow. It does not take up any space. A shadow is not matter.

But most things that you see *are* matter. Even things that are very small, like an ant, a speck of dust, or a grain of sand. And even things that are much smaller than that.

Nearly the smallest things in the world are *atoms.* Atoms are like tiny building blocks that make up all the matter in the world. Atoms are everywhere. They are in everything you touch. A lamppost is made of atoms. A dog biscuit is made of atoms. *You* are made of atoms.

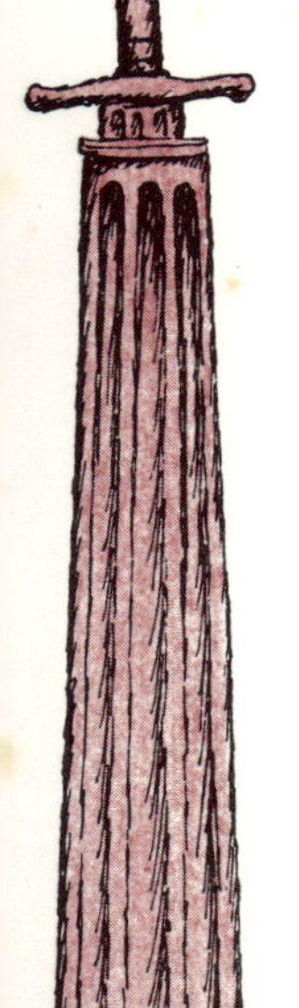

Atoms are smaller than anything you have ever seen. In fact, nobody has ever seen an atom all by itself. When you look at a desk or a cat or a toothbrush, you are seeing countless numbers of atoms all together. But seeing a single atom all alone is just about impossible. Even the most powerful microscope could not help you to see an atom.

There are about a hundred different kinds of atoms that make up all the materials and objects around you. Things can be made up of just one kind of atom, or of several different kinds. When a material is made up of just one kind of atom, it is called an *element.* Iron is an element. Lead is an element. Gold is an element. There are about one hundred elements.

An element remains what it is no matter how little of it you have—even just an atom's worth. If you took an iron nail and sawed it into pieces, all the pieces would still be iron. Even if you could cut up

the nail until you had nothing left but a single atom, that atom would be iron.

Atoms of different elements can be put together in various ways. When they are, they make substances called *compounds.*

The salt that you use on your table is a compound. It is made up of an element called sodium, and another one called chlorine. But that does not mean that if you pulled apart a grain of salt, you would find sodium atoms and chlorine atoms all mixed up every which way. Instead, you would find each atom of sodium joined up with an atom of chlorine in something called a *molecule.*

A molecule is the smallest thing you could have

that would still be salt. It is not like a sodium atom or a chlorine atom, but something altogether different. If you broke up a salt molecule, then you would not have salt any more, but just an atom of sodium and an atom of chlorine.

All compounds are made up of molecules. Usually the molecules are not at all like the atoms of which they are made. When the atoms join up, they make something completely new.

Sodium and chlorine are both poisons, for example. Even a very small amount of either of these two elements by itself could harm you. But when they join to become table salt, they make a compound that you can put on your food.

SODIUM

CHLORINE

SALT

Many of the materials you see and use are compounds made up of two or more elements. Often the same elements can join together in different ways to make compounds that are not at all alike.

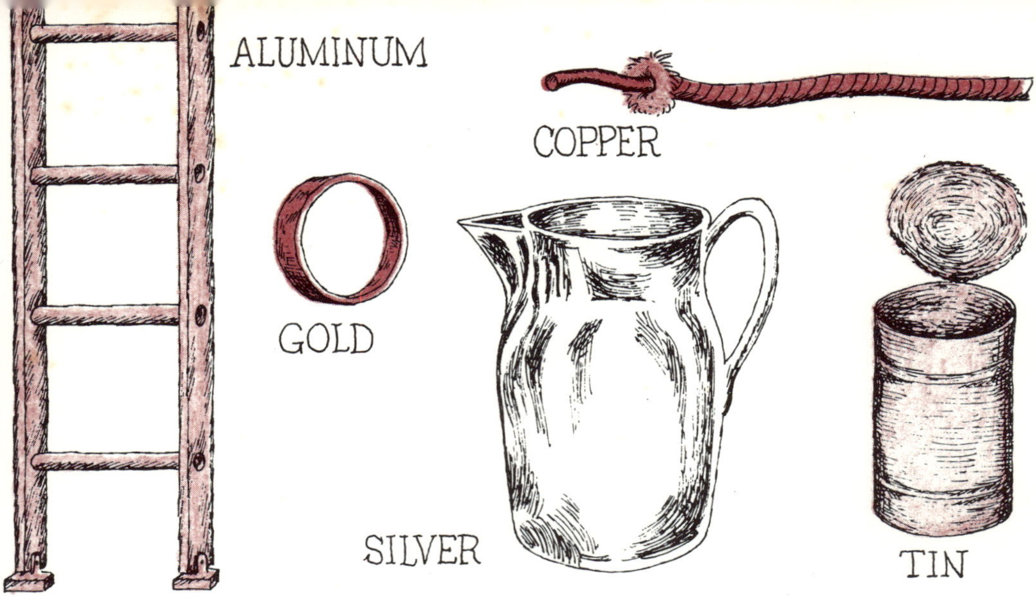

That is why there are so many different materials in the world. The atoms of the one-hundred-or-so elements can join up in different ways to make molecules of so many different compounds that probably no one has ever counted them all.

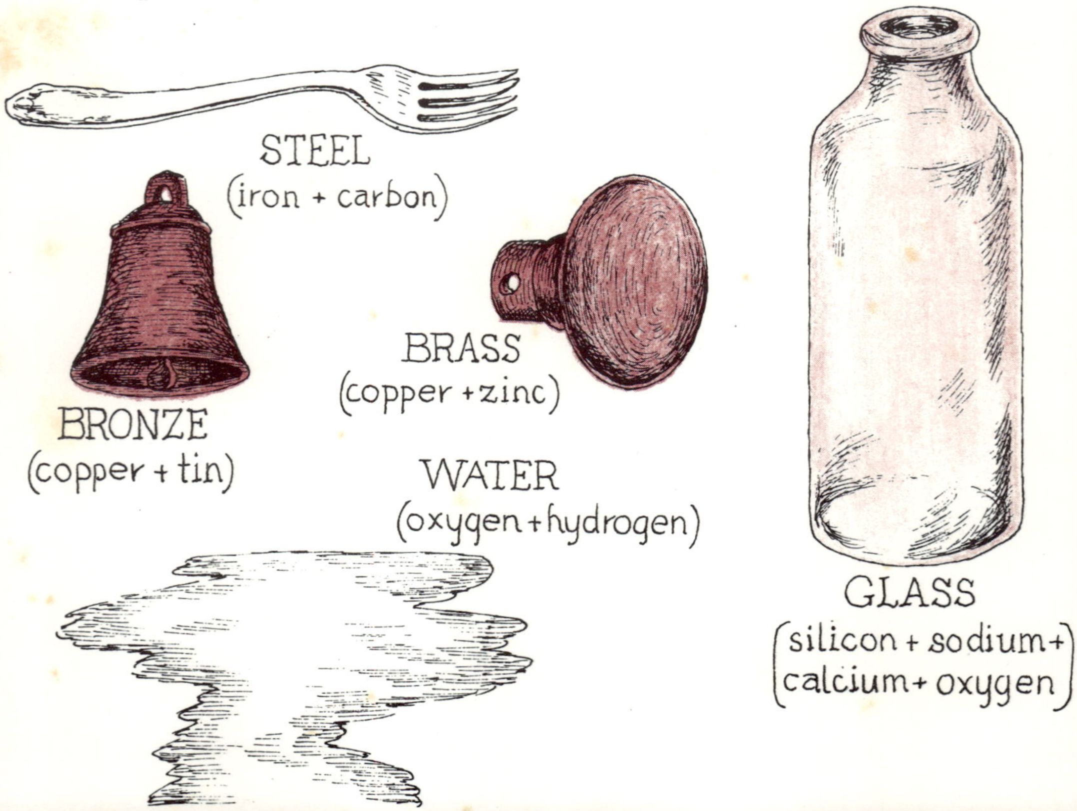

Three

Two Milks and a Glass of Cookie, Please

Because all kinds of matter are made up of atoms, you could say that they are all alike in that way. But still, different kinds of matter can be very, very different.

For example, a wooden block is matter, and water is matter. But they are not very much alike. You don't even talk about them the same way. If you wanted to float a wooden block in water, you might

say "I need some water and a block," not "a water and some block."

Why is that?

Blocks are easy to count. You can count 1, 2, 3 blocks in a box. Can you count how many waters there are in a bottle?

Blocks can be taken out of a box one by one and put on the floor. Can you take waters out of a bottle one by one?

A block and water are both matter. But they are different kinds of matter.

A block is a solid thing. Water is a liquid.

They both take up space. A block takes up space in a box or on the floor. Water takes up space in a bottle or in a glass.

A block and water cannot take up the same space at the same time. If you push a block into a glass of water, some of the water gets pushed aside by the block. So where the block is, there is no water.

A block and water can both be touched. But a different thing happens when you touch them. When you touch water, your finger goes right through it. Your finger does not go through a block. If a block has a hole in the middle, your finger can go through the hole. But it is not going through the block. It is going through a space in the block.

Blocks are easy to count because each block is a separate thing. If you line up some blocks on the floor, you can see where one block ends and another one begins. If you pour some water into a dish, you have just one big pool of water.

A block has a shape. It may be a rectangle, or it may be square. It may be a triangle, or some other shape. But each block has *some* shape.

Does water have a shape? When it is in a bottle, it has a bottle shape. When you pour it into a glass, it has the same shape as the glass.

What happens if you pour it into a glass that has a different shape?

What happens if you put a block into a glass? Does it get a new shape?

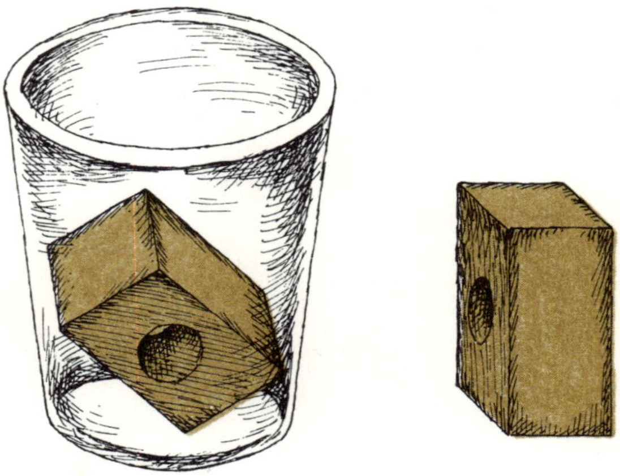

A block is a solid thing. It has a shape that is always the same. If you cut a block up into pieces, the pieces can have different shapes. But each piece will keep its shape, too. When you drop a block on the floor, it makes a thump. Sometimes it might break.

Water is a liquid. It does not have a shape of its own. Water always has the same shape as the bottle, dish, or glass it is in. If you drop some water on the floor, it does not break or make a thump. It makes a splash, and forms a wide, flat puddle on the floor. The water spreads out as far as it can.

A block does not spread out if you drop it on the floor. It stays in one small place. Even if it breaks, the pieces usually do not scatter very far.

How else is a block different from water?

Four

A Solid Is Hard, Soft, Smooth, Rough...

All around you there are solids.

Books, trees, squirrels, earmuffs are solid things. Like blocks, they come in different shapes; but each has a shape that is always the same. They can be counted. You can say 1, 2, 3 books, trees, squirrels, earmuffs.

They take up space. If a book is in a certain space, you cannot put a carrot in the same space, unless you move the book. If you put the carrot on top of the book, it is in a different space. If you put a mouse into a teacup, it is not in the same space as the teacup. It is in a space *in* the teacup.

Solid things cannot go into each other. You could not push a walnut into a table. You are able to push a pencil through a piece of paper because the pencil makes a hole in the paper that it can go through.

Is a mouse in a teacup inside the solid matter that makes up the teacup?

Even though they are all alike in some ways, solids can be very different from each other, too. A wooden block is very hard. You would need a knife or a saw to change its shape. But you don't need a knife to change the shape of a piece of clay. Just squeeze the soft clay in your hand, and it will get a new shape.

You can change the shape of a pillow by squeezing, too. But if you leave the pillow alone for a while, it will get its original shape back. Will a stick of clay do that?

Solid things come in every shape you can imagine, and some you probably can't. They come in all different sizes and colors. Some solid things have spaces in them, like the hole in a doughnut or the drawer spaces in a desk. Other things are all one solid piece, like a marble or a baseball bat.

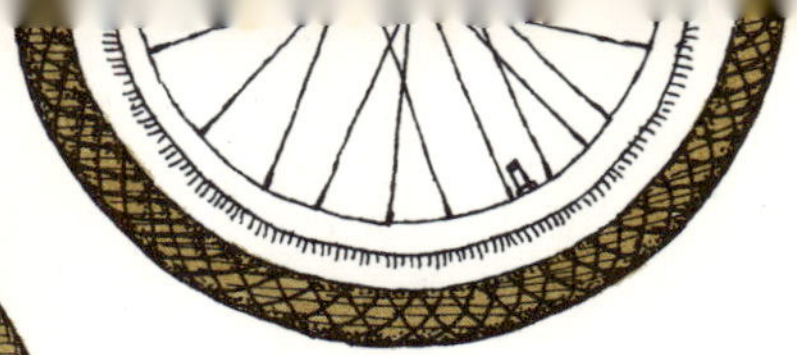

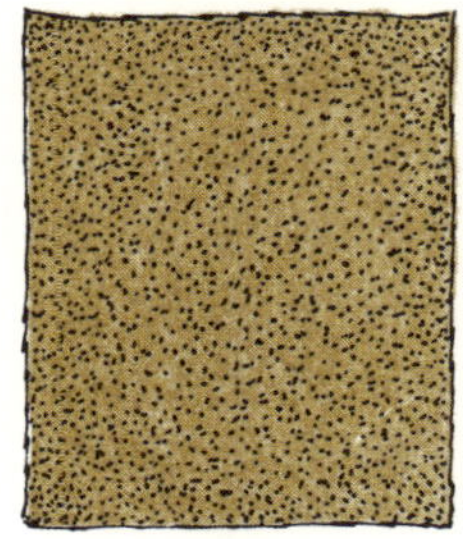

Solid things may feel very smooth, like a mirror; or rough, like sandpaper. They may feel silky, like a cat; or sort of bumpy, like a bicycle tire. Solids can be as shiny as a new brass belt buckle, or as dull as an old iron fence. They may be sharp, like the point of a pencil; or blunt, like the eraser end.

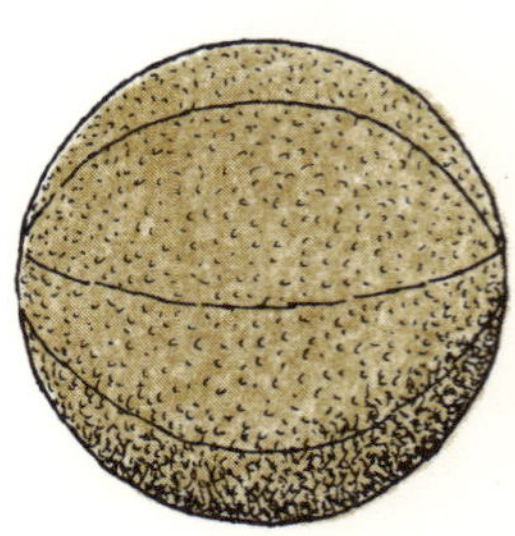

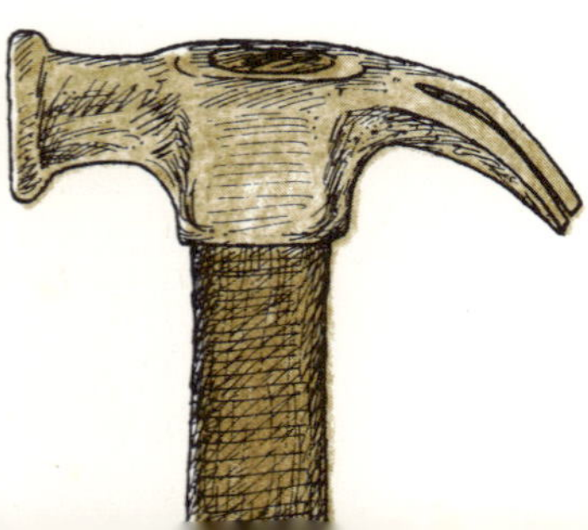

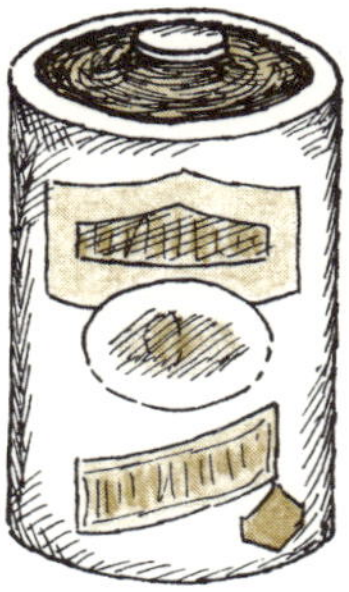
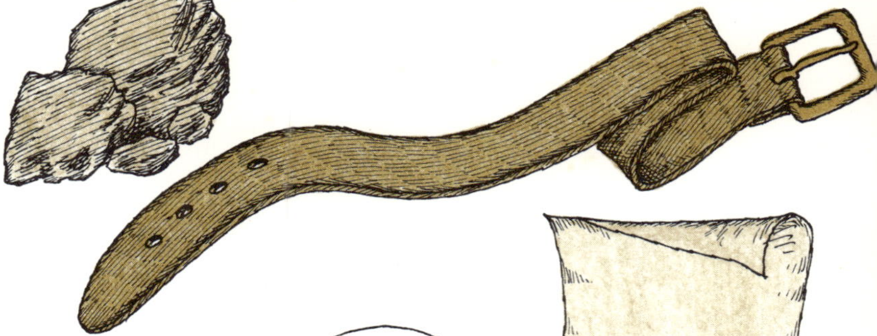

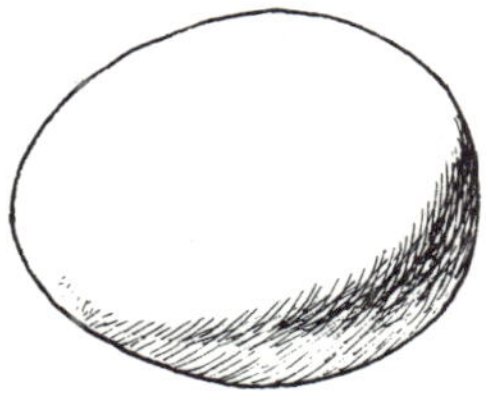

How else might solids look and feel? Look all around you. Can you find things that are thick and things that are thin? Things that break when you drop them, and things that bounce? Things that can bend, and things that break when you try to bend them? Things that are alive, and things that are not?

Things, things, solid things all around you. How many ways can you find to tell about them?

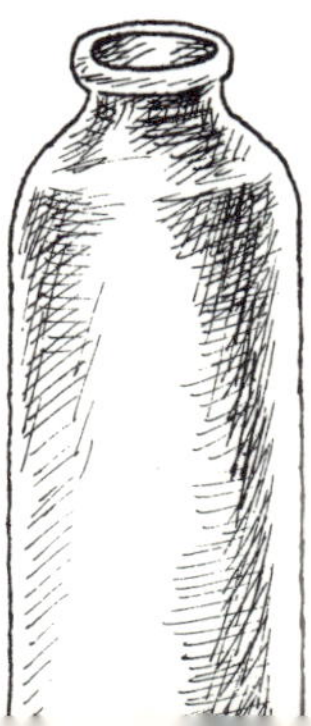

Five

Thick as Water, Clear as Molasses

All around you there are liquids.

Milk, orange juice, lemonade, cough syrup are liquids. Like water, each has a shape that is always changing. When you pour cough syrup from a bottle into a spoon, the cough syrup loses its bottle shape, and gets a teaspoon shape.

Liquids can always be poured from one thing to another. Can you pour solids? Can you pour books, squirrels, or earmuffs?

One liquid can be poured into another liquid. If you pour water into orange juice, the orange juice will take up more space. It will take up as much space as the orange juice and the water did separately, before you mixed them.

Solid things can be put into a liquid. A raisin will go into a glass of water. If you put a lot of raisins into a glass of water, you will see the water rise in the glass, because the raisins are taking up some of the space inside. The water and the raisins will take up as much space in the glass as they both did when they were separate.

Liquids cannot be counted like solids. You cannot count 1, 2, 3 waters or orange juices. How can you say how much liquid you have?

You can pour water into a glass until it fills up the glass, and say it is a glassful. Or you might pour orange juice into a pitcher, and say it is a pitcherful. You could pour cough syrup into a teaspoon, and say it is a teaspoonful. That is one way of saying how much liquid you have.

Liquids usually feel wet, like water. But some liquids feel sort of sticky, like molasses. Water flows quickly when you pour it. But when you pour molasses–well, it's slow as molasses!

Water is a thin liquid; molasses is a thick one. Paint in a can is thick, too. Turpentine is thin, like water. When a painter adds turpentine to a can of paint, he makes the paint thinner. Then the paint flows more easily from his brush.

Liquids are different in other ways, too. They come in all different colors—bright ones and dull ones.

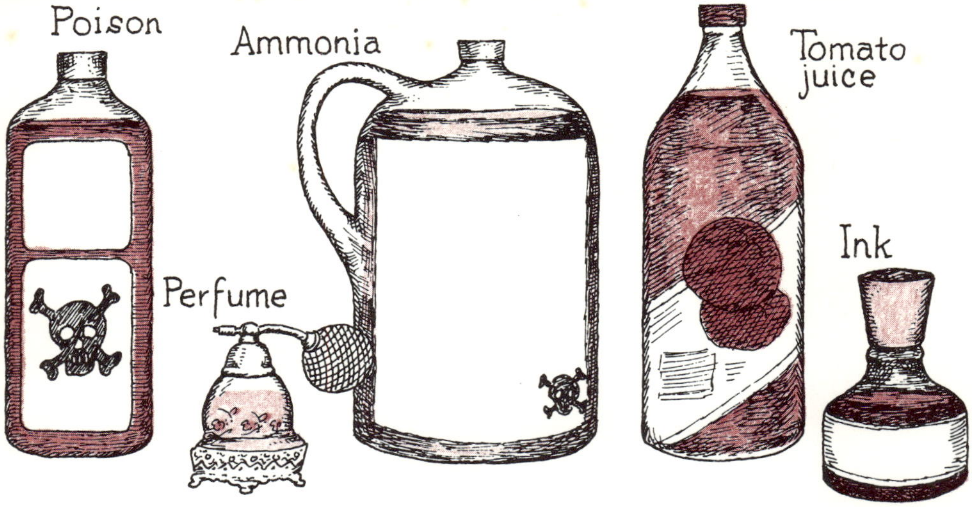

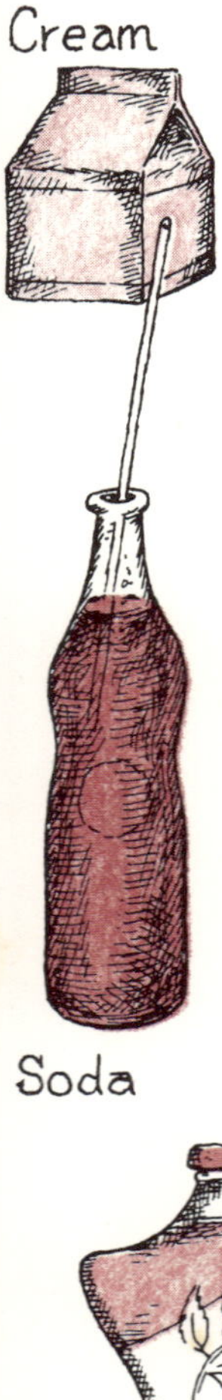

Some liquids, like water, you can see right through; others you can't. Some liquids smell good, like perfume. Others smell awful, like ammonia or gasoline. And some liquids have no smell at all.

How else can liquids be different from each other? Can you find liquids that are good to drink, and liquids that are not? Liquids that can burn, and liquids that can put out fires? Liquids that are good for cleaning things, and liquids that make things dirty?

How many different things can you find out about the liquids all around you?

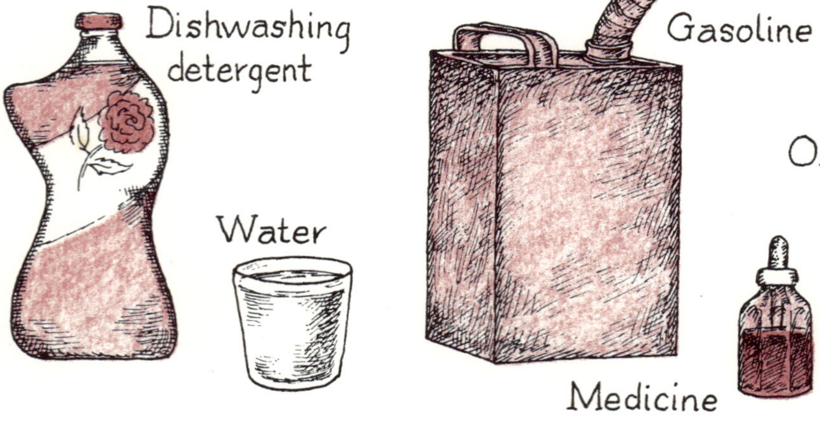

Six

Some Heavy Reading

All around you there is solid and liquid matter. How can you tell how much there is?

One way is to see how much space it takes up. (How can you do that?) Another way is to see how heavy it is. When a grocer weighs things on a scale, he is seeing how heavy they are.

Some things are very heavy. They have a lot of weight.

Some things are very light. They have only a little weight.

But everything that is matter has *some* weight.

A giggle does not have any weight. A giggle is not matter.

A cookie has weight. It is solid matter.

Milk has weight. It is liquid matter.

It is easy to see how much weight a cookie has. Just put it on a scale if you have one, or can find one to use. If it is a very large cookie, it might weigh an ounce. Then it would take 16 of the cookies to make a pound. A pound is 16 ounces. But most cookies would probably weigh less than an ounce. If it took five cookies to make an ounce, then there would be 80 cookies to a pound.

That is why, when you buy a box of cookies, the label tells how many ounces or pounds of cookies are in the box, instead of telling exactly how many cookies there are. It is quicker to just put cookies in the box until you have a pound, than to count out 80 cookies.

Many food packages tell how much of a solid they have in them, rather than how many. Would you like to count the grains of rice that go into a box? It is easier just to say that they weigh a pound!

Weighing liquids is a little different.

You cannot pour milk onto a scale. It will run right off. But still, you can use a scale to weigh some milk. You can weigh a glass on a scale and see how heavy the glass is. Then you can pour some milk into the glass, and see how much heavier the glass gets when the milk is in it. The extra weight is the weight of the milk.

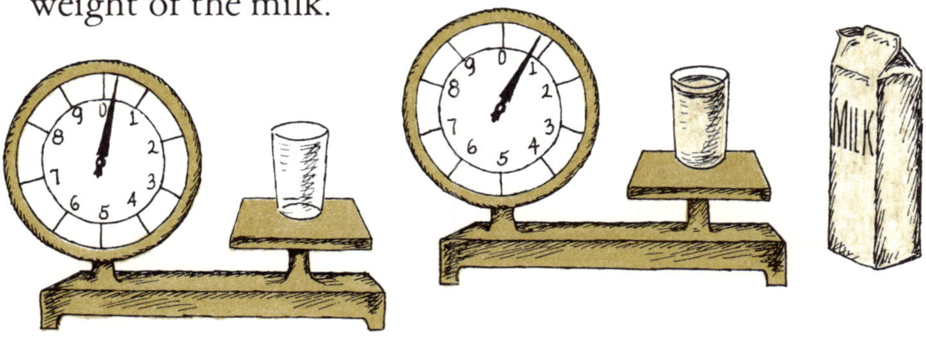

Even though liquids can be weighed on a scale, they are not usually sold by weight. Since you have to put a liquid into a container anyway, it is easier just to say how big the container is. Then you needn't bother weighing it.

Liquids like milk are usually sold by the quart or pint. Liquids that you use less of, like poster paint or shampoo, may be sold by the ounce. There are 32 liquid ounces to a quart, and 16 to a pint.

Liquid ounces are not the same as weight ounces. They measure space, not weight. A liquid that took up one liquid-ounce-worth of space might have more or less than one ounce of weight.

The more you have of a thing, the more it weighs. A large box of rice will weigh more than a small one. A big, ripe watermelon will weigh more than a little one that has just started to grow. A quart of milk will weigh more than a pint.

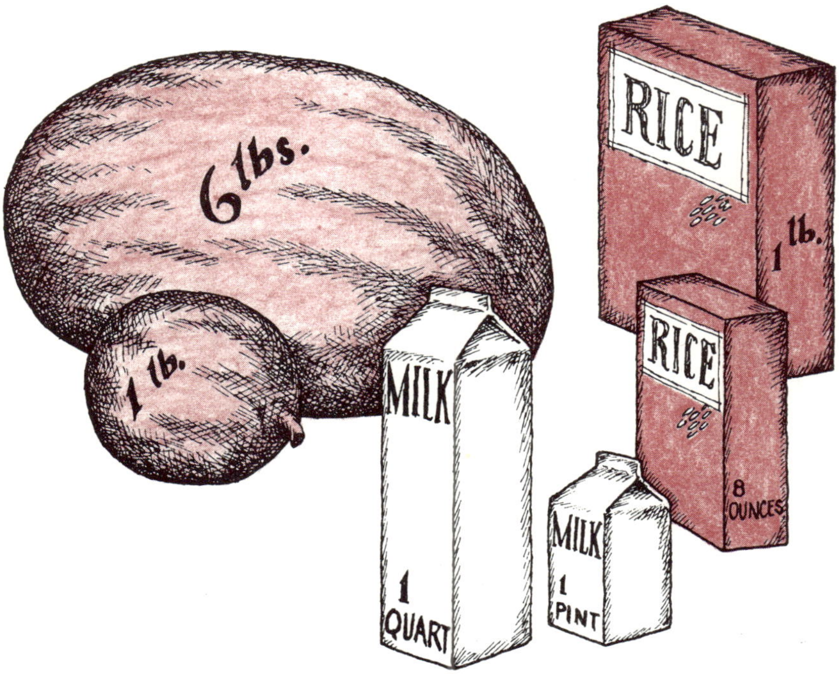

But if you have things of two different kinds, it is not always so easy to tell which will be heavier. Some things have more weight for their size than others.

A box full of books would weigh more than a same-size box full of marshmallows. A baseball weighs more than an orange. A brick weighs more than a whole pillowful of feathers!

Find some things in your home or school to weigh. You can weigh heavy things on a bathroom scale, a nurse's scale, or a grocery scale. You can weigh light things on a kitchen scale, a laboratory scale, or a postage scale.

Which is heavier, a hammer or a toy truck? A glass of orange juice or a glass of soda pop? A pitcher of milk or a pussycat?

Are you as heavy as your best friend?

Seven

Liquid or Solid?

Usually it is easy to tell the difference between a solid and a liquid. Diving into a pool full of water would not be the same as diving into a pool full of bricks!

But sometimes it is not so easy to tell what something is.

Pour some sugar into a glass. The sugar pours into the glass something like the way water does. Is sugar a liquid?

Drop a teaspoon into the sugar. Does the spoon fall through the sugar to the bottom of the glass, the way it would fall through water?

When you pour sugar into a glass, the sugar does not stay in one small place. It spreads out to cover the bottom of the glass. Water does the same thing. But when you fill a glass with water, the top of the water is flat. Is the top of the sugar flat?

Sugar can be poured like a liquid. But things cannot go into sugar as easily as they can go into a liquid like water. Sugar spreads out something like the way water does, but not as much. Could sugar be a thick liquid?

Dip the tip of a teaspoon into the sugar, so that just a very small amount of sugar gets onto the spoon. Shake the sugar around to make it spread out. Can you see separate pieces of sugar on the spoon? You will see them if you look carefully. They are very tiny. It will help if you have a magnifying glass to make them look bigger.

Push a tiny piece of sugar around on the spoon with a pin. The piece of sugar has a shape. It does not change its shape when you push it around. You can pick it up with the pin and put it on your finger. Can you do that with water?

Sugar is made of many tiny, solid pieces. When you pour sugar, the solid pieces fall all over each other into a pile. It looks *something* like what happens when you pour water. But the sugar is solid, not liquid.

If you have some marbles, put them into a bag or box and then pour them out into a bowl. Do you see how the marbles fall all over each other, and spread out in the bowl? That is very much like what happens when you pour sugar.

What other solids can you pour something like the way you pour a liquid? Try sand, flour, talcum powder. How can you tell that they are not really liquids?

What happens if you drop them into a pail or bowl on the floor? Do they make a thump, and perhaps break? Or do they make a splash and a puddle? Or do they do something in between?

Eight

Thump Goes the Water

Solids and liquids can change. A liquid can become a solid. A solid can become a liquid.

Put some water into an ice-cube tray, and put the tray into the freezer part of your refrigerator. Take the tray out after a few hours, when the water has become frozen. Take an ice cube out of the tray, and put it on a dish.

Does the frozen water spread out in the dish? Can you pour it from the dish, or does it just fall off the dish in one lump? Does it make a splash and a puddle if you drop it in the sink? Or does it just make a thump?

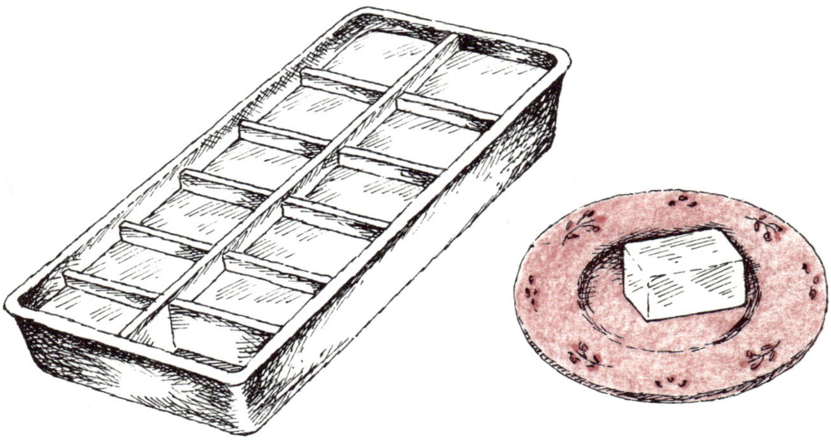

The water is not a liquid any more. When you froze the water, it became a solid—ice. What happens if you let the solid ice stand in a dish on your table? Watch carefully, and you will see the water slowly become a liquid again. You will see it spread out over the dish.

Inside the freezer it was very cold. The cold made the water change from its liquid form, or *state,* to its solid state, ice. The room you are in is not so cold as the freezer. In the warm room, the solid ice turns to liquid water again. Ice is just solid water.

What other liquids can you turn into solids by putting them into the freezer? Try orange juice, lemonade, milk. Will they all become liquids again if you let them stand on a table after freezing them? Try it and see.

Some things that are usually solids can become liquids if you make them warmer. Watch what happens when a grown-up puts some butter or shortening into a frying pan on the stove. You will see how the heat from the burner makes the butter melt. You will see the solid butter become a liquid that spreads out over the bottom of the pan.

What do you think would happen to the liquid butter if the pan were taken off the stove to cool? Find out!

Nine
It's a Solid . . . a Liquid . . . a Gas!

When you take an ice cube out of the refrigerator and let it stand in a dish, it slowly changes to water. What happens if you let the water stand in the dish?

If you leave the water in the dish for a day or two, and look at it from time to time, you will notice something strange. The water seems to just disappear!

How can that happen?

To understand that, you have to think about something much smaller than an ice cube–smaller even than a tiny piece of an ice cube. You have to think about something no bigger than a molecule.

Now a molecule is a very small thing to think about, so it's easier to think about a lot of molecules all together. An ice cube is made up of countless molecules of water.

Molecules are always moving around. But in an ice cube, the molecules are sort of pushed together, and don't have room to move very far. You can imagine that they are something like people in a crowded elevator.

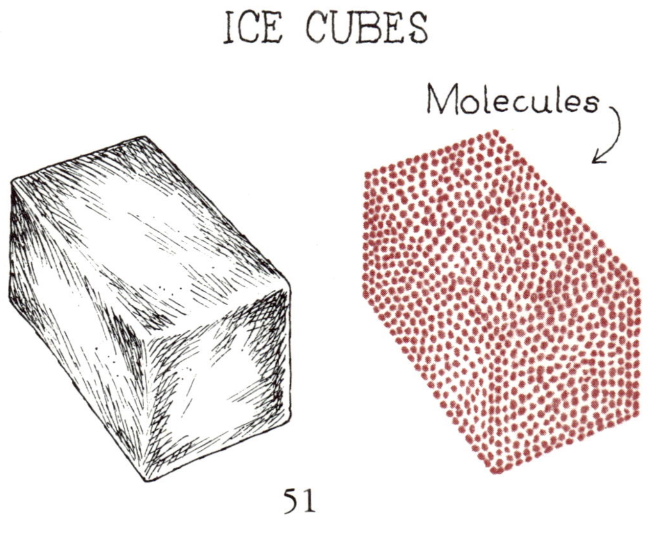

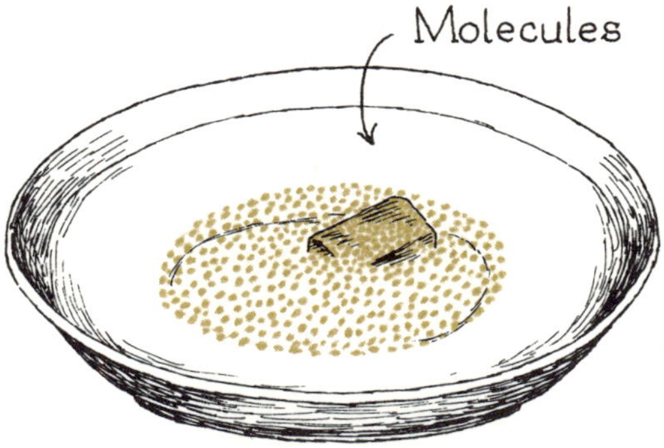

When an ice cube melts, the molecules become more free to move–something like people getting out of the elevator when the door opens. The molecules can slip and slide and fall all over each other, like the marbles that you poured into a dish.

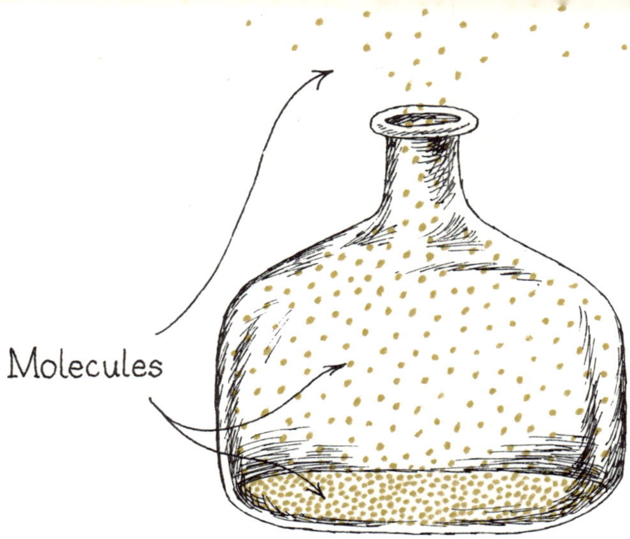

When the ice cube is all melted, the molecules have so much room to move around in, and start moving so fast, that some of them jump right out of the dish. The molecules keep on doing that until there are none left at all. Heating the water makes this happen faster.

Once the water molecules leave the dish, the water is in a new state. The liquid water has changed to water *vapor.* (It has *evaporated.*)

Water vapor is a gas. When water becomes a gas, the molecules have lots of room to move up, down, and sideways. They can travel very far apart from each other. If they are in a container, they can fill up the whole container—no matter how big it is—and take its shape. They can leave the container through the tiniest opening—even at the top—and spread out through the air.

Water vapor in the air can become liquid water again if it is cooled. Have you ever seen tiny drops of water on the outside of an ice-cold glass of lemonade? The water did not come from inside the glass. Water vapor in the air became cooled when it touched the cold glass, and changed back to liquid water on the side of the glass.

Air is also a gas, or really several gases all mixed together. An important one is oxygen. You need oxygen to breathe. Oxygen is also one of the two elements that make up water. The other one is hydrogen, which is also a gas. But there is very little hydrogen in the air. When oxygen gas and hydrogen gas come together, they form liquid water.

Now that you know something about the way molecules behave, you can probably figure out why things can go into liquids and gases, but not into

solids. When you get into a bathtub full of water, or walk into a room full of air, your body pushes the molecules of the air or water aside, and makes a space for itself.

But the molecules in a solid are too jammed together for that. You cannot walk into a wall because the molecules in the wall won't move aside to let you in!

Ten

About Gases and Other Matters

Why are gases in this book about matter—about things that take up space, things that have weight, things you can touch?

Gases are in this book because they do, they have, and you can.

Gases are matter.

When you take a glass out of the cupboard, it has air in it. The air gets pushed out when you fill the glass with milk. Milk and air cannot both fill up the glass at the same time. If the air could not get out, the milk could not get in.

Stuff a tissue into the bottom of a glass so it won't fall out when you turn the glass upside down. Then push the upside-down glass straight down into a sink or a bucket full of water. (Don't smash the glass against the bottom.)

Does the tissue get wet? Some of the air in the glass couldn't get out when you pushed the glass into the water. So the water couldn't get in to wet the tissue.

Whenever you blow up a balloon, you are showing that gases take up space. The air you blow into the balloon makes the balloon get bigger and bigger.

When you let go of the balloon without tying a knot, most of the air comes out again. If you let the air out of a balloon over some small bits of paper, you will see the paper get pushed away by the air. But the air does not make a thump when it comes out of the balloon, or even a splash. It just makes a *whoosh!*

A blown-up balloon has more air inside it than a balloon that is not blown up. Does it weigh more? If you have a balance, or a scale that can measure very light things, you will see that it does. Like all matter, air has weight.

Can you touch air? You can, even though you may not feel it all the time. But when you are out on a windy day, or when you stand near a fan or an air conditioner, you can feel the air touching you. A wind outdoors or a breeze from a fan is moving air. You can feel it on your body, and you can feel whether it is warm or cool.

Many elements and compounds are gases. Some, like air, you cannot see; others you can. Some gases help keep you alive, like oxygen. Other gases could make you ill if you breathed them. Some gases can burn, or can help things burn. Other gases can be used to fight fires.

Can you think of any gases that you use? Helium is a very light gas that makes balloons sail high in the air.

Carbon dioxide gas makes the bubbles that tickle your nose when you drink soda pop. Some dentists give you "laughing" gas to make the drilling hurt less. And all the time you are surrounded by the air that you breathe.

Gases are *really* all around you!

Everywhere you look there is matter—even matter that you cannot see. It may be solid, or liquid, or gas. Some things may be more than one.

Soda pop is a liquid and a gas. A bowl of vegetable soup is both solid and liquid. A sponge is a solid thing with lots of tiny spaces in it that hold air. When you use a sponge to mop up a spill, then it contains a liquid, too.

Can you think of any kind of matter that is *not* a solid, a liquid, or a gas?

Index